Materials

BY
CHARIS MATHER

©This edition published 2025.
First published in 2023.
BookLife Publishing Ltd.
King's Lynn, Norfolk
PE30 4LS, UK

All rights reserved.
Printed in India.

A catalogue record for this book is available from the British Library.

HB ISBN: 978-1-80155-830-3
PB ISBN: 978-1-80155-863-1

Written by:
Charis Mather

Edited by:
William Anthony

Designed by:
Drue Rintoul

All facts, statistics, web addresses and URLs in this book were verified as valid and accurate at time of writing. No responsibility for any changes to external websites or references can be accepted by either the author or publisher.

PHOTO CREDITS

Cover – Victoruler, My Life Graphic, ArtParnyuk, IhorL, Oleksandr Rybitskiy, dordek, Chones, Tom Wang. 4–5 – Tatiana Gordievskaia, Hatchapong Palurtchaivong, Vecton, Prostock-studio. 6–7 – Photographee.eu, Africa Studio, New Africa. 8–9 – alinabuphoto, Liudmila Pereginskaya, Purino. 10–11 – Dmytro Zinkevych, StockImageFactory.com. 12–13 – Stepan Kapl, CGi Heart, FabrikaSimf. 14–15 – Virunja, Ilina93, pikselstock, chomplearn. 16–17 – Bezuglova Evgeniia, Cavan-Images. 18–19 – sirichai chinprayoon, Iam_Anuphone, mar_pet. 20–21 – kzww, Maples Images. 22–23 – Gelpi, Pixel-Shot, Jiri Hera, Steve Collender, Crisher. All images are courtesy of Shutterstock.com, unless otherwise specified. With thanks to Getty Images, Thinkstock Photo and iStockphoto.

Contents

PAGE 4	Think like a Scientist
PAGE 6	Materials
PAGE 8	Glass
PAGE 10	Plastic
PAGE 12	Metal
PAGE 14	Wood
PAGE 16	Fabric
PAGE 18	Changing Shapes
PAGE 20	In the Real World
PAGE 22	Act like a Scientist
PAGE 24	Glossary and Index

Words that look like this can be found in the glossary on page 24.

Think LIKE A Scientist

What do you see around you? You might have questions about some of what you see. Scientists are people who find out things about the world around us.

When scientists do not know something, they make a guess and check to see if they are right. Scientists who learn about materials do tests to help people choose the best ones for different jobs.

Materials

Everything in the world is made of a material. Materials have properties, which are the things that make objects different from others. The properties objects have make them good at some things and bad at others.

What is a good material for a spoon? A wooden spoon would work well. A spoon made of cloth would not work. Cloth does not have the right properties to do a spoon's job.

Cloth

Wood

Wood is hard and keeps its shape, unlike cloth.

Glass

Glass is used for things such as windows, glasses and bottles. Glass is hard and does not bend. Most glass is also transparent. This means it is see-through and lets light through.

Broken glass is sharp and dangerous. **NEVER** touch it!

Glass is also waterproof, which means that water cannot get through it. Glass is a good material for fish tanks. It can keep water in and be seen through.

Plastic

Plastic is another waterproof material. Plastic is different from glass because it can be hard or bendy. Plastic is used to make lots of things, such as bags, bottles, toys and cups.

Plastic is sometimes transparent and sometimes not. Something that you cannot see through is opaque. Plastic is not as heavy as glass. It is also strong, which makes it a good material for safety helmets.

Metal

Metal shares some properties of glass and plastic. It also has its own properties. Metal is strong, opaque and waterproof. There are lots of different types of metal. Some are heavy and some are not.

Aluminium is not heavy.

Gold is heavy.

Metal is shiny. You can also bend and heat up metal without breaking it. Metal is used to make lots of things, such as spoons, pans and paperclips.

Wood

Wood is a material that is used to make doors, sheds, chairs and many other things. Wood is hard, but not as hard as glass. You cannot see through wood because it is opaque.

Someone who makes things out of wood is called a carpenter.

Wood is different from glass and metal because it can burn up. This means that it can be set on fire. When wood burns, it turns into ash, which is a powder that is left over.

Ash

Fabric

There are lots of different types of fabric, and they each have different properties. Fabric is not hard and can be bent and twisted. Some fabrics are soft and some are rough.

Clothes, blankets and curtains are all made of fabric. Some fabric is made by weaving together thin strands called fibres. Fabric can be many different colours.

Changing Shapes

Materials do not always stay the same shape. Changes can happen when materials are heated up, bent, squashed, cut or twisted. When glass, metal and plastic are hot, they can be made into different shapes.

Glass can be made into different shapes by using fire.

Being able to change a material's shape is very useful. Wood can be cut and carved into different shapes for building. Metal can be very thin and detailed, or it can be thick and strong.

Both of these chains are made of metal.

IN THE Real World

This house is made of wood.

Knowing how materials are different is very important in the real world. Our houses need to be made with strong, waterproof materials. Do you know what material your house is made of?

Knowing about materials also helps our planet. Some materials, such as plastic, are not good for the environment. Instead of throwing these materials away after one use, we should reuse and recycle them.

Plastic, metal, glass and paper can all be recycled.

Act Like a Scientist

What materials do you have in your house? Do you think any of them would be good materials to use as boats? Have a think about what properties a boat needs.

A boat should be waterproof and able to float.

Collect some objects made of different materials with a grown-up. You could try:

- Bottle lids
- Marbles
- Lollypop sticks
- Keys
- Hair bobbles

Put each object in a bowl of water.
Which ones are waterproof? Do any float?

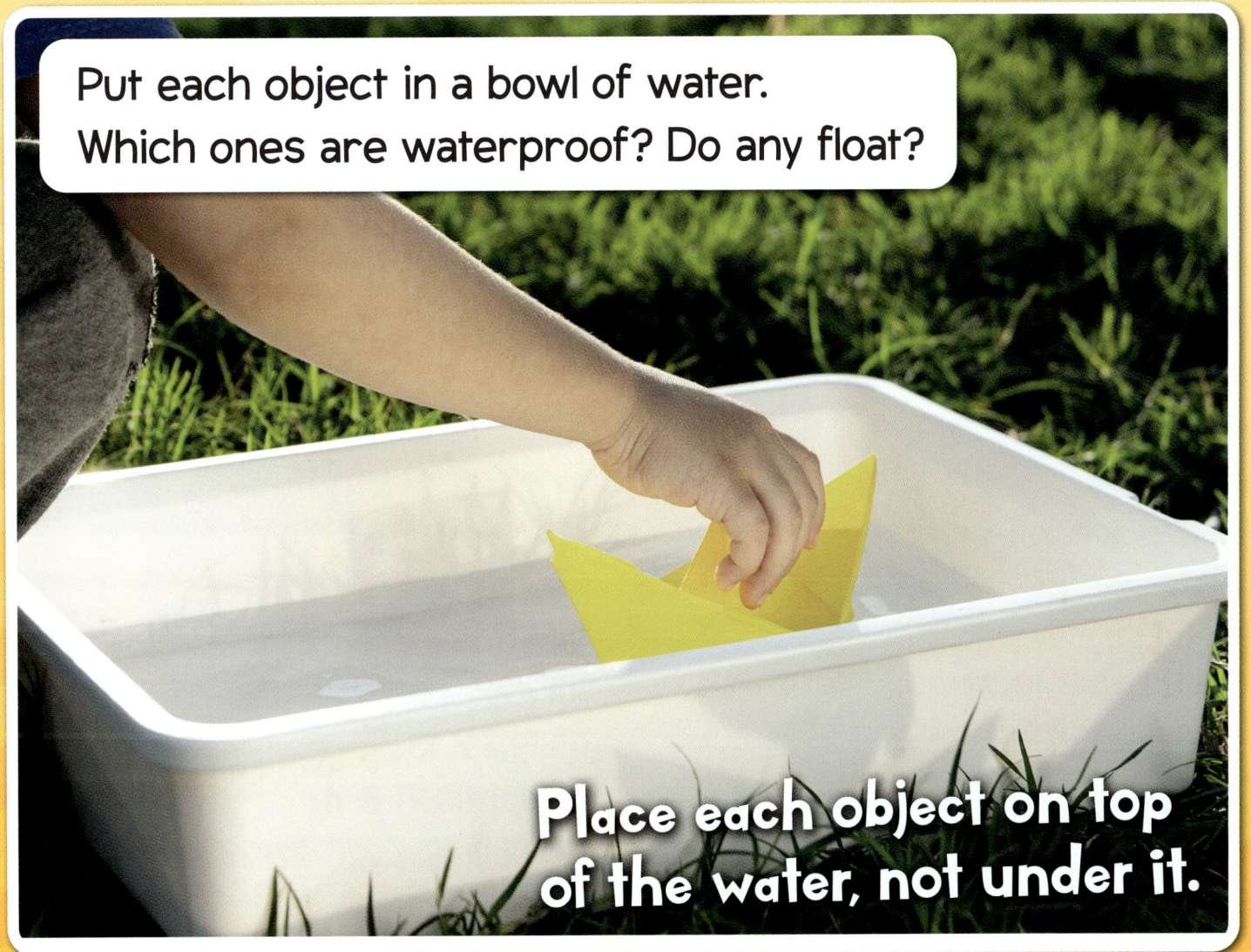

Place each object on top of the water, not under it.

Glossary

CARVED	cut into a shape
CHAINS	rows of metal rings that are linked together
ENVIRONMENT	the natural world
FLOAT	to rest on the top of water, instead of going under it
RECYCLE	use again to make something else
ROUGH	uneven and not smooth
SAFETY HELMETS	hard plastic hats made to keep people's heads safe
STRONG	difficult to break
SQUASHED	flattened or crushed
WEAVING	criss-crossing long threads to make a usable material

Index

ASH 15
BOTTLES 8, 10, 23
BURNING 15
HARD 7–8, 10, 14, 16
HEAT 13, 18
OPAQUE 11–12, 14

SHAPES 7, 18–19
SOFT 16
SPOONS 7, 13
TRANSPARENT 8, 11
WATER 9–10, 12, 20, 22–23